I0756099

THIS BOOK BELONGS TO:

THE WONDERFUL WORLD OF PANGOLINS

MIMI JONES

Dedicated to all who love pangolins.

ISBN 978-1-970416-05-3

www.joeysavestheday.com

Mimi Books™ Publishing

A Mimi Book

Pangolins are the only mammals covered in hard, protective scales. These scales help keep them safe from predators.
Mammal

Strong

Their scales are made of keratin, the same material as our fingernails. This makes them strong but lightweight.

Protects

When scared, a pangolin curls into a tight ball. This shape protects its soft belly.

Stretch

A pangolin's tongue can stretch longer than its whole body. It helps them reach deep into insect nests.

Their tongue is attached near their pelvis instead of their throat. This gives it extra length and flexibility.

Flexibility

Pangolins don't have any teeth. They swallow their food whole.

Teeth

They eat mostly ants and termites. That makes them insect-eaters, or insectivores.
Ants
Termites
Insectivores

One pangolin can eat up to 70 million insects in a single year. This helps keep insect populations under control.

70 Million

Pangolins use their strong claws to break open ant and termite mounds. Their claws are sharp and powerful.

Powerful

Their saliva is sticky like glue. It helps them catch lots of insects at once.

Saliva

Pangolins have poor eyesight. They rely on their excellent sense of smell to find food.

Eyesight

Nostrils

They can close their nostrils and ears while eating. This keeps insects from crawling inside.

There are eight different species of pangolins in the world. Each one has its own special features. Four species live in Asia, and four live in Africa. They all share similar habits.

Some pangolins live high up in trees. These species are great climbers.

Climbers

Tree-dwelling pangolins have long, gripping tails. Their tails help them balance and hang from branches.

Balance

Ground pangolins live on the forest floor. They dig burrows to sleep and hide in.

Forest Floor

A pangolin's burrow can be up to 11 feet deep. These tunnels keep them cool and safe.
Burrow

Baby pangolins are called pangopups. They are born with soft, bendable scales.

Pangopup

Pangolins are mostly active at night. This makes them nocturnal animals.

Nocturnal

Pangolins sometimes walk on their hind legs. They do this when carrying food or moving quickly.

Hind Legs

Their scales make a soft clattering sound when they move. It's one of the ways you can hear them coming.

Clatter

Insects

Pangolins help the environment by eating insects. This keeps ant and termite numbers balanced.

Aerate

When they dig for food, they help loosen and aerate the soil. This makes the ground healthier for plants.

Pangolins can live in forests, grasslands, and savannas. They adapt well to different habitats.

FOREST

GRASSLAND

SAVANNA

They communicate by leaving scent markings. These scents help them share information with other pangolins.

Scent

Grind

Pangolins have strong stomach muscles that grind their food. Their stomach acts like a natural grinder.

Smelly

Pangolins can release a smelly odor from glands near their tail. This scent helps scare away predators.

Even though they look like armadillos, they are more closely related to cats and dogs. Their appearance can be surprising.

Appearance

Pangolins can live up to 20 years in the wild. Their secretive nature makes them hard to study.
Wild

Ancient

Pangolins have existed for millions of years. They are ancient animals with a long history.

Learning

Scientists are still learning new things about pangolins every year. These shy animals keep many mysteries hidden.

Pangolins are one of the most trafficked mammals in the world. Many people are working hard to protect them.

Endangered

All eight species of pangolins are threatened or endangered. They need our help to survive.

Count the pangolins.

Thank you for exploring The Wonderful World of Pangolins with me. I hope you learned something new and enjoyed discovering these gentle, scale-covered creatures who curl into the cutest little balls.

If you liked this book, please consider leaving a review. It helps other families discover it too.

See you in the next adventure!

Check out these other interesting books in the Wonderful World of series!

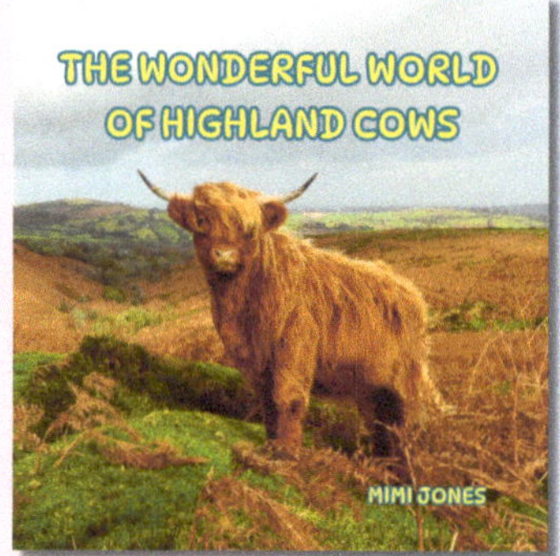

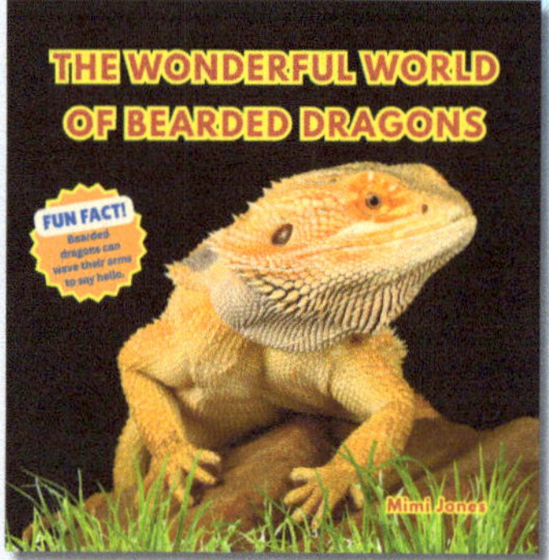

www.mimibooks.com

www.ingramcontent.com/pod-product-compliance
Lightning Source LLC
LaVergne TN
LVHW070158110826
845147LV00002B/438

* 9 7 8 1 9 7 0 4 1 6 0 5 3 *